Wilson Fox

The Artificial Production of Tubercle in the Lower Animals

A Lecture Delivered at the Royal College of Physicians, May 15, 1868

Wilson Fox

The Artificial Production of Tubercle in the Lower Animals
A Lecture Delivered at the Royal College of Physicians, May 15, 1868

ISBN/EAN: 9783337075217

Printed in Europe, USA, Canada, Australia, Japan

Cover: Foto ©berggeist007 / pixelio.de

More available books at **www.hansebooks.com**

THE ARTIFICIAL PRODUCTION OF TUBERCLE

IN THE LOWER ANIMALS.

A LECTURE

DELIVERED AT THE ROYAL COLLEGE OF PHYSICIANS,

MAY 15, 1868.

BY WILSON FOX, M.D. F.R.C.P.

HOLME PROFESSOR OF CLINICAL MEDICINE AT UNIVERSITY COLLEGE, LONDON.

London:

MACMILLAN AND CO.

1868.

LONDON:

R. CLAY, SON, AND TAYLOR, PRINTERS,

BREAD STREET HILL.

PREFACE.

As the inquiry into the origin of Tubercle by the agency of direct irritation or by septic matter appears to me to be one of the most important advances which have been recently made in the pathology of the disease, I have thought it desirable, at the suggestion of several friends, to publish my observations on the subject in a separate form. The contents of this lecture are almost identical with the verbatim reports which appeared in the spring in the *British Medical Journal* and the *Lancet*. I have, however, supplied a few further details from fresh observation, and have in some places slightly expanded the argument where it appeared to require further elucidation. I am, also, by this means, enabled to lay before the profession in greater number the illustrations which appear to me to prove the tubercular nature of the affection thus artificially produced, and I therefore trust that this object will appear to justify the republication of matter which has so recently been brought under their notice.

I must, in conclusion, express my sincere thanks to Mr. H. B. Tuson for the high artistic power which he has bestowed on the elucidation of the disease, and also to Mr. Ford for the great skill and care with which he has executed the plates.

22 B, CAVENDISH SQUARE,
Sept. 1868.

DESCRIPTION OF PLATES.

PLATE I.

Fig. 1.—Lungs of Guinea-pig, containing tubercles, grey and semi-transparent at the margins, and in some places slightly opaque in their centres.

Fig. 2.—Enlarged Spleen of Guinea-pig, showing scattered grey granulations, which in some places are agglomerated into groups; in other parts, groups of granulations are seen of opaque yellow colour.

Fig. 3.—Portion of another Spleen when the granulations are larger and more scattered.

Fig. 4.—Liver of Guinea-pig, showing large tracts infiltrated with grey granulations, passing in many places into a more opaque yellow condition. Larger and more isolated opaque whitish spots are also seen scattered through the tissue.

Fig. 5.—Axillary Lymphatics of Guinea-pig, showing cheesy spots.

Fig. 6.—Subcutaneous Granulations and Cheesy Masses near seat of injury, in rabbit inoculated with tubercle. (These were identical in appearance with those described in the guinea-pig.) The masses are seen to be composed of agglomerated granulations. Smaller groups of these are seen at variable distances from the larger masses. These latter are greyer and less cheesy than the larger masses.

Fig. 7.—A Cord of Indurated Tissue, partly cheesy, extending between a lymphatic gland and a cheesy granulation. (Guinea-pig.)

Fig. 8.—Lobular Pneumonia in a Rabbit, the subject of Laryngo-tracheitis. The infiltration of the pulmonary air-vesicles forms a marked contrast to the granulations in the rabbit.

Figs. 9, 10.—Lungs of Rabbit with Pyæmic Spots, contrasting with the granulations in the guinea-pig.

PLATE II.

Fig. 1. *A, B.*—Tubercular Growth in Sheath of Bronchi. (Guinea-pig.)

A × 460 diam. shows (*a a a*) section of bronchial tube at point of bifurcation. The upper part is marked by elastic fibres; the lower, by cartilage cells. *b b b* represents the growth of tubercle in its sheath, which is seen to be proceeding by a multiplication of cells and nuclei, partly round, partly ovate and fusiform. These, at a little distance, are passing into the walls of the alveoli, which are thickened by the growth; the outlines of the alveoli being still maintained. A few enlarged epithelial cells are seen within the alveoli. The vessels of the alveoli so implicated are for the most part obliterated.

B × 700 diam. From lower part of *A*. (*a*) Sheath of bronchus. (*b*) Growth of tubercle by round and ovoid cells. (*c*) Epithelium enlarged and separating. (*d*) A growth of fusiform cells which also are seen passing in strings and rows between the capillaries; of which a good example is observed at (*e*).

Fig. 2.—Growth of Tubercle in Perivascular Sheath of Pulmonary Artery. × 460 diam. (reduced). (Guinea-pig.)

(*a a*). An artery at a point of bifurcation, the section being carried obliquely through the plane of both branches.

(*b b*). Multiplication of cells in the sheath, external to the muscular coat. In both branches a dense agglomeration of these cells is seen in some parts, marked in one branch by (*c*). The growth is seen extending into the walls of the surrounding air-vesicles, the capillaries of which are impervious to injection.

(*d*). Enlarged and pigmented epithelial cell.

Fig. 3.—Transverse Section of a Pulmonary Artery, surrounded by a growth of Tubercle. The outlines of the perivascular sheath are here obliterated, and the structure of the wall has become indistinct. (a) artery; (b) masses of nucleated cells and nuclei surrounding the vessel; (c) fusiform cells mingled with the growth; (d d) lines of fusiform cells surrounding the capillaries; (e) a capillary obliterated; and (f) another capillary obliterated, with enlarged nuclei in its walls. × 700 diam.

Fig. 4.—Shows the gradual Thickening of the Walls of the Alveoli surrounding a Tubercular Granulation. The gradual obliteration of the lumen of the air-vesicles is distinctly seen. × 100 diam.; binocular arrangement. (Guinea-pig.)

Fig. 5.—Enlargement of Nuclei of Capillaries of Lung preceding their obliterations. × 700 diam. (Guinea-pig.)

Fig. 6.—Enlargement of Nuclei of Capillaries of Lung. **Failure of injection.** × 700 diam. (Guinea-pig.)

Fig. 7.—Growth of Masses of Nucleated Cells around Capillaries in Walls of Alveoli of Lung at margin of a mass of Tubercle. At (a a) a denser mass of tubercle, with capillaries partially obliterated. × 700 diam. (Guinea-pig.)

Fig. 8.—Growth of Fusiform Cells around Capillaries of Air-vesicles. × 700 diam. (Guinea-pig.)

PLATE III.

Fig. 1.—Omentum of Guinea-pig. Tubercle around a small artery (a), which bifurcates into two branches (b b). The tubercle is seen to consist of a mass of nucleated cells, and of nuclei around which no outer cell wall is visible. These which are more closely agglomerated around the vessel become more scattered at greater distances. The meshes of the omentum become gradually obliterated by the growth. Some large nucleated cells are seen among the growth. × 700 diam.

Fig. 2.—Omentum of Guinea-pig. Masses of tubercles situated on the vessels (a b), a small artery and attendant veins. The growth of cells can be seen in places extending along the vessels for some distance from the larger masses. The extension into the neighbouring tissue of the omentum can also be observed. × 100 diam. (reduced).

Fig. 3.—Omentum of Guinea-pig. Small mass of tubercle, having no apparent connexion with vessels. The thickening and filling of the meshes of the omentum with cells and nuclei can be seen here, as in Fig. 1. × 700 diam.

Fig. 4.—Tubercle of Liver (Guinea-pig). (a a a a) Acini of liver. (b b) Bile ducts, marked by columnar epithelium. (c c) Growth of tubercle, which extends in all directions between the acini, pressing them aside. (d d d) Spots where the tubercular growth is passing between the cells of the acini separating these. (e e) Isolated liver cells, left unchanged amid the growth of the tubercle. × 100 diam.

Fig. 5. A, B.—Isolated portions of the Tubercular Growth in the Liver. × 700 diam.
A. Shows nucleated cells imbedded in the meshes of a fibrous network.
B. A mass of nuclei forming a string or obliterated tube.

Fig. 6.—Portion of Subcutaneous Tissue amid Granulations near seat of injury, showing a multiplication of oval, round, and fusiform cells in the fibrous tissues. At (a a a) is a dense mass of nuclei, forming an irregular string or cord. × 700 diam.

Fig. 7.—Isolated Cells, from margin of Fig. 6. The larger cells measure 1-890th × 1-400th of an inch. Fusiform cells are seen in process of division. × 700 diam.

Fig. 8.—Section of a Subcutaneous Granulation, near seat of injury. Shows nucleated cells imbedded in the meshes of a fibrous network. × 700 diam.

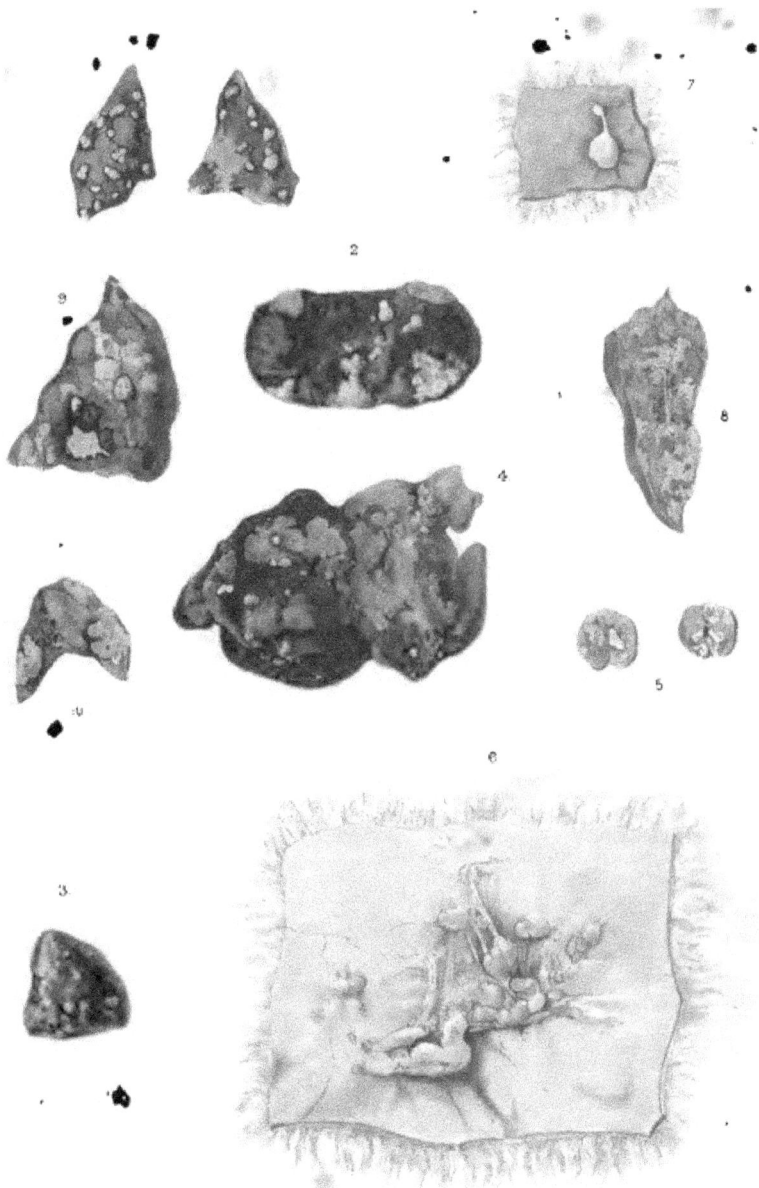

.

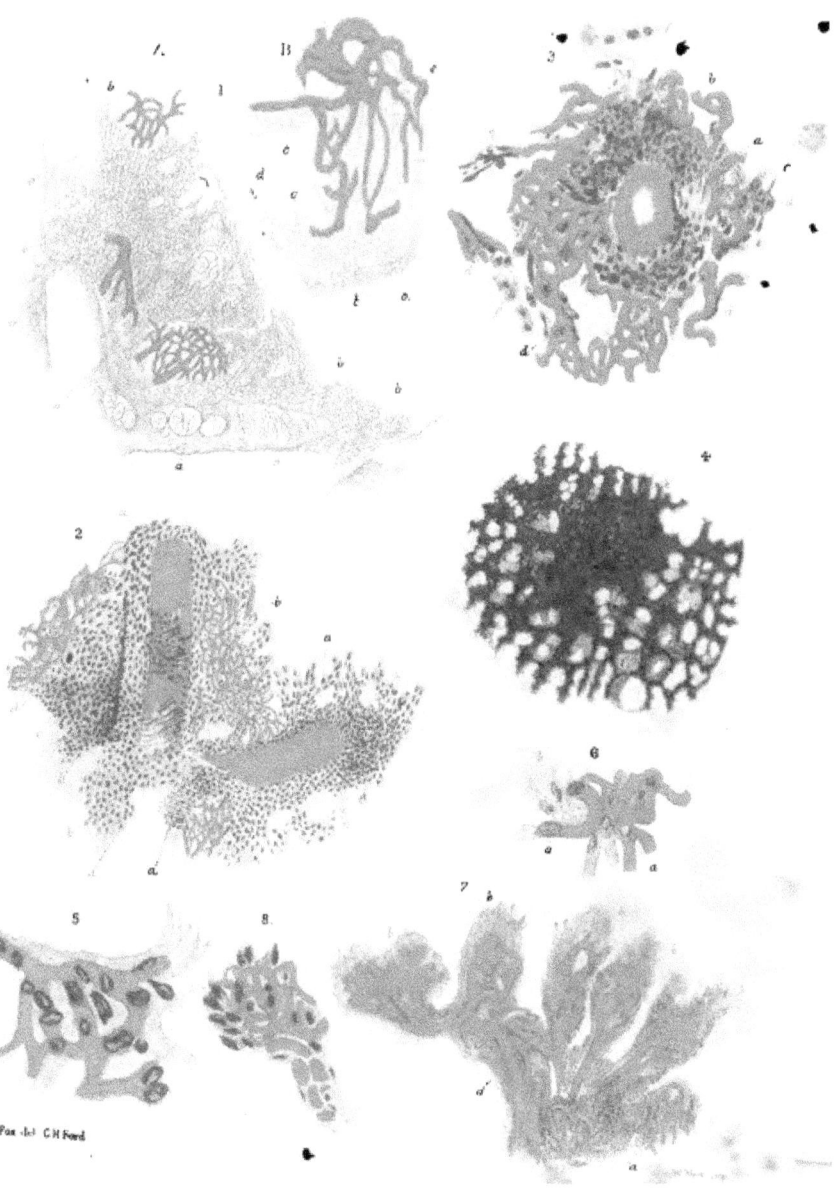

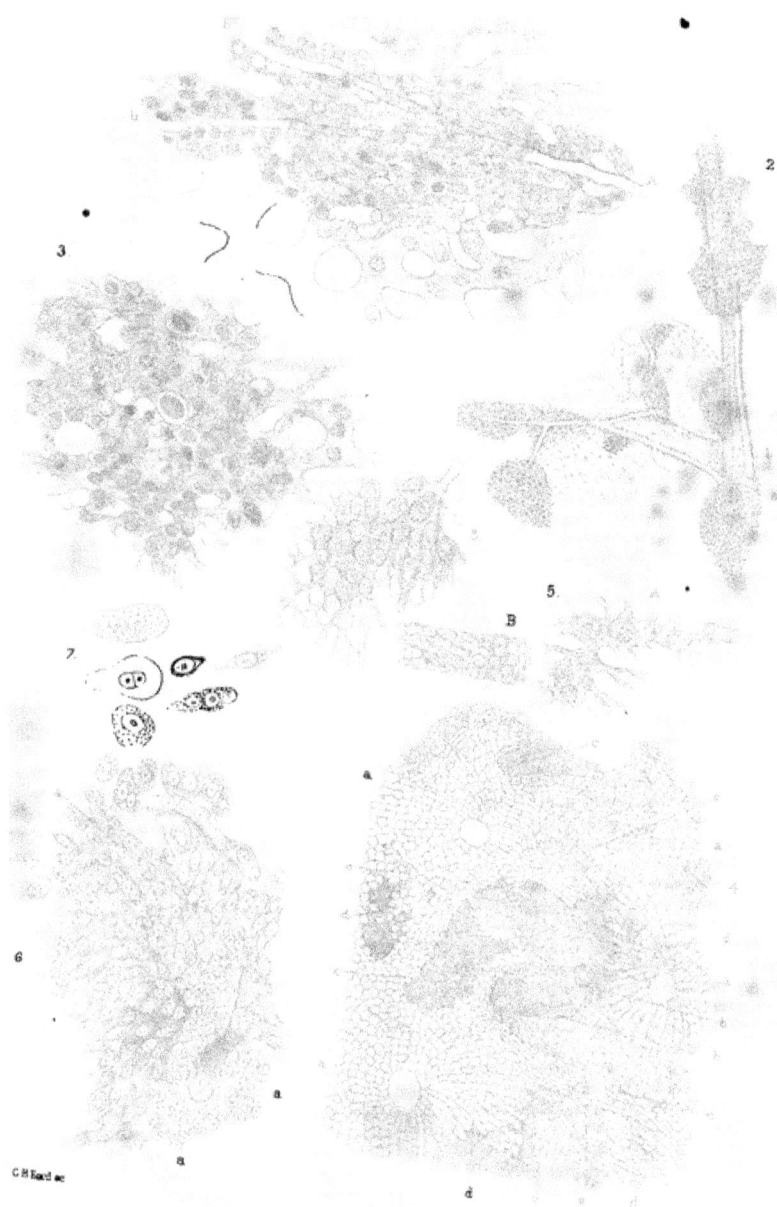

2

3

5.

B

7

6

C.H.Kard sc

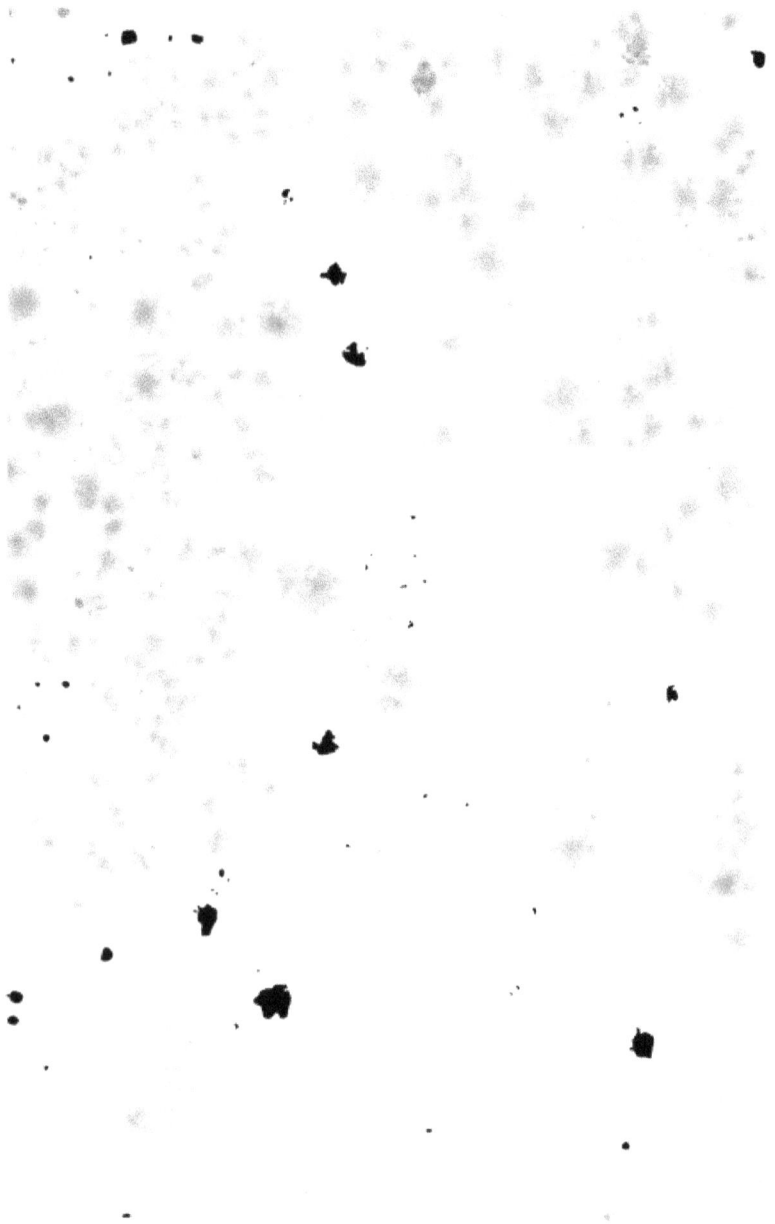

A LECTURE

THE ARTIFICIAL PRODUCTION OF TUBERCLE IN THE LOWER ANIMALS.

Mr. PRESIDENT AND GENTLEMEN,

THE nature and relations of tubercle have of late received such able and elaborate illustration in this College, both by Dr. Andrew Clark in the Croonian Lectures, and by Dr. Southey in the Gulstonian Lectures, of last year, that I should not have presumed again to call attention here to this subject, had not the question received within the past year some new and, as it appears to me, most important additions, through experiments conducted by myself and others on the artificial production of the disease, which I cannot but deem to be deserving of the serious consideration of the able and independent observers whom this College numbers among its members.

The genesis of tubercle, although long involved in profound obscurity, has nevertheless been constantly and repeatedly a subject of experiment by some of the most illustrious pathologists of the past and present century.

Thus, Baron[1] stated that the first Dr. Jenner had observed that rabbits could be made tuberculous by submitting them to a special kind of diet. That the appearances thus produced were due, however, only to hydatids is seen by an illustration of Sir R. Carswell's,[2] where the affection of the liver is shown now to be due to entozoic disease. It must, however, be remembered that Baron, and probably Jenner, attributed to hydatids an important part in the production of tubercles.

---

[1] Inquiry illustrating the Nature of Tuber-
culated Accretions, 1829, pp. 96, 97.

[2] Illustrations of the Elementary Forms of
Disease. "Tubercle," pl. ii. fig. 6.

B

Others, as Kortum, Hebréard, Lepelletier,[1] Guersant, Alibert, and Richeraud,[2] have all attempted the inoculation of scrofulous matter from man upon animals, but without results; and the only successful series of this nature, among those of earlier date, were the experiments of Erdt[3] upon horses, where, after the inoculation of scrofulous pus in the nostril of the animal, the submaxillary glands enlarged, and nodules of a doubtful character were found in the lungs.

Laennec[4] also considered that an induration in his own hand, arising from a puncture contaminated by the pus of a carious spine, might possibly be of a tuberculous nature.

The question, however, received no further elucidation in this direction until M. Villemin, led by induction to believe that tuberculosis was a specific and contagious disease, performed a series of successful experiments with the inoculation of tubercle upon rabbits, which he communicated to the French Academy of Medicine in 1865, and added the details of a further series in 1866.[5] M. Villemin selected several pairs of rabbits from the same litter. He inoculated one with tubercle, and left the other. He placed them all under similar conditions, and found that those inoculated became almost invariably tuberculous, while the others escaped. These experiments were too numerous to leave any doubt of the effect produced. M. Villemin also experimented with other substances, such as the materials of anthrax, of phlegmonous abscess, of pneumonia, cancer, typhoid, and the stools of cholera, all on rabbits; but with none of these could he produce the same effects.

In 1866, M. Lebert also communicated to the French Academy a series of experiments confirmatory of M. Villemin's results.[6]

M. Villemin's statements were submitted to a committee of the Academy, and reported upon, in July 1867, by M. Colin, who, in the main, confirmed M. Villemin's observations. To some facts contained in this report, on the inoculation of other animals, I shall have hereafter to refer. In February 1866, MM. Hérard and Cornil[7] confirmed M. Villemin; but stated that grey tubercle alone was capable of reproducing tubercle when inoculated, and that this power was not possessed by yellow tubercle.

In April 1867, Mr. Simon, President of the Pathological Society, laid before that body the results of a considerable number of experiments, in which, by the same means, he had succeeded in reproducing tubercle in the

[1] Virchow, Die Krankhaften Geschwulste, ii. 724.
[2] Hérard and Cornil, La Phthisie Pulmonaire, 552.
[3] Virchow, loc. cit.
[4] Traité d'Auscultation Médiate, p. 221.
[5] Bull. Acad. Méd., xxxi. xxxii.
[6] Ibid. xxxii.
[7] La Phthisie Pulmonaire.

rabbit.[1] His results were submitted to a committee of the Society, who, after examination, expressed their conviction that the disease thus produced was similar to what is ordinarily received as tubercle in man. At that meeting, and subsequently at the Croonian Lectures, Dr. Andrew Clark[2] made the important statement, that he had succeeded in producing the same results as could be obtained by the inoculation of tubercle by using other non-tuberculous pathological products, but expressed some doubts whether the disease so produced corresponded to tubercle in man; and Mr. Barwell also stated that he had observed what appeared to be tubercles in the lungs of rabbits who had suffered from injuries to their bones. A little later in the same year M. Villemin's researches were confirmed by Professor Hoffman and M. Genondet; and Dr. Marcet, in a paper read before the Royal Medical and Chirurgical Society, proposed to use the sputa of patients as a means of diagnosis of tubercular disease.[3]

These statements appeared to me to involve questions of the utmost importance, and I determined to take an early opportunity to carry out a series of experiments as to the effects of different morbid materials introduced under the skin of rabbits and guinea-pigs in the same manner as M. Villemin had conducted his inquiry with tubercle.

I commenced my experiments in July 1867, but found, at the time when I was beginning, that a distinguished Fellow of this College, Dr. Sanderson, had shortly before instituted a series of experiments with the same object. I did not know what was the precise line of his investigation. The question also appeared to me to be involved in so much obscurity and doubt, seeing that, with the exception of two cases of Dr. Andrew Clark, the experiments hitherto made seemed to show that tubercle alone when inoculated was capable of producing either tubercle or a disease resembling tubercle,— a conclusion which must modify the whole of our views of the pathological nature of this disease,—that I considered myself justified in continuing the independent inquiry which I had marked out.

It has proved that, within a few weeks of one another, perfectly independent and without any knowledge of each other's results, Dr. Sanderson and I have arrived at the same conclusion. A similar result has also been recently obtained by Dr. Waldenburg.[4] Dr. Sanderson and I have both found that non-tubercular substances introduced under the skin of guinea-pigs

[1] Path. Soc. Trans., vol. xvii.

[2] Medical Times and Gazette, 1867, pp. 366, 429.

[3] Med.-Chir. Trans., vol. l.

[4] Allg. Medicin. Cent. Zeitung, Dec. 14, 1867. Dr. Waldenburg's paper only came into my hands after my own experiments were completed.

are followed by the production of a disease which we consider tuberculous; and the results of my own observation, which I have now the honour to lay before you, confirm Dr. Sanderson's views, which he published at the Pathological Society a few weeks since.

The only other experiments which here require allusion have been some recently made by Professor Lebert,[1] on the introduction of matters into the circulation; and to these I shall have again to refer.

As regards my own experiments, I have inoculated with various materials 117 guinea-pigs, and 12 rabbits. Of the 117 guinea-pigs, 58 have proved tubercular, 6 have yielded doubtful results, and no effect was produced in 53.

In order to save time, and prevent a wearisome repetition of figures, I have tabulated the results which I have obtained.

I wish to explain that I have placed those cases only in the category of tubercle in which, in addition to local effects, or to the implication of the neighbouring lymphatic glands, *three or more* of the following organs have been found to be simultaneously affected; viz. the lungs, the bronchial glands, the liver, spleen, omentum, mesentery, and intestines. When this has been the case, I have tabulated the result as successful; when only one or two, as doubtful; when none, as failure. In the next columns, I have stated the duration of life, and in many cases whether the animals died or were killed. In another column I have placed what I consider a very interesting point; viz., the number of times in which there has been what I shall explain as a local production of tubercle. Unfortunately my notes do not in all cases contain a statement of this fact.

---

[1] Virchow's Archiv, vol. xl.

RESULTS OF EXPERIMENTS.

No. of Inoculations	Tubercle produced	Doubtful	Failure	Duration of life Tubercle. Days. *		Duration of life. Systems. Days. *		Local production of Tubercle			Number of internal organs affected.			*Remarks. Where no letter is placed (as these columns) the animal was not killed. The letter k signifies killed; d, died.)
				Min.	Max.	Max.	Min.	Present	N stand	Absent	Average	Min.	Max.	
8	6	0	2	92	5	8	2	4	2	2	5	6	5	
1	1	0	0	45	...	...	...	1	...	...	4			
2	2	0	0	84	63	...	...	2	...	...	5½	7	4	
4	4	...	...	95	35	...	...	4	...	...	7½	8	5	
9	2	...	...	50	29	...	...	2	...	...	6½	8"	5	*Cholera.
2	2	...	...	71	46	...	...	2	...	...	5	6	4	
2	0	0	2	...	...	15"	...	0	...	2	...	...	...	*Frost.
13	11	...	2	...	...	...	...	...	...	...	...	...	...	
2	0	0	2	...	...	16	2	0	...	2	...	...	...	
2	0	0	2	...	...	89 k	88 k	0	...	2	...	...	...	
2	1	0	1	26	...	20 b	...	0	1	1	2	...	...	
6	1	0	5	...	...	...	...	0	...	...	...	...	...	
1	0	0	1	...	...	54 k	...	0	...	1	...	...	...	
2	0	0	2	...	...	175 k	97 k	0	...	2	...	...	...	
6	0	0	2	...	...	50 k	17	0	...	2	...	...	...	
5	3	0	2	158 k	41	179 k	6	2	...	3	6½	7	5	Spleen only.
2	0	1	1	56	...	87 k	...	...	0	2	...	1"	0	
2	1	0	1	117 k	...	172 k	...	...	...	2	...	6	...	
1	1	...	...	56	...	...	...	1	...	...	...	4	...	
1	1	...	...	59	...	...	...	1	...	...	6	5	...	
1	1	...	...	60	...	...	...	1	...	...	...	...	...	
12	7	1	9	...	...	...	...	...	...	...	...	...	...	
3	0	0	3	...	...	113 k	4 b	...	...	5	...	...	...	
1	0	0	1	...	...	...	112 k	...	...	4	...	...	...	
3	1	...	2	65	...	158 k	6	0	...	3	...	7	4	
2	2	...	...	88	50	...	...	2	...	...	5	7	3	
2	1	...	1"	95	...	79"	...	1	...	1	...	7	...	*Frost.
11	4	0	7	...	...	...	...	...	...	...	...	...	...	
2	2	0	0	166 k	49	...	...	2	...	...	4	4	...	
2	0	1"	1	...	...	100 k	6	0	...	2	...	1"	...	*A few granula-
1	1	0	0	41	...	...	...	1	...	...	...	7	...	tions in lungs.
1	...	1"	...	...	...	75 b	...	0	...	1	0"	...	...	*Great emaciation
3	3	...	...	109	80	...	...	3	...	...	6½	7	4	(and dropsy.)
9	6	2	1	...	...	...	...	...	...	...	...	...	...	
5	5	0	0	68	29	...	...	4	1	...	4½	6	4	
2	2	0	0	115 k	75	...	...	2	...	...	6½	7	6	
1	1	0	0	53	...	...	...	1	...	...	...	4	...	
1	1	0	0	31	...	...	...	0	...	1	...	4	...	
2	3	...	...	60	59	...	...	3	...	...	4½	7	0	
12	12	0	0	...	...	...	...	...	...	...	...	...	...	
3	0	0	3	...	...	113 k	61 k	0	...	3	...	...	...	
4	0	0	4	...	...	100 k	...	0	...	4	...	...	...	
3	1	1"	6	119 k	113 k	57 k	28 k	0	...	3	3	4	2"	*Only bronchial
4	4	6	0	86 k	78 k	...	...	1	...	3	4½	7	3	glands and spleen.
6	0	1"	4	...	...	71 k	16 b	0	0	4	0	0	1"	*Lungs doubtful.
3	1	0	2	44 b	...	73 k	...	0	0	3	4	...	...	
4	1	1"	2	51 k	...	51 k	29 x	1	0	3	5	...	1"	*Lungs.
2	0	0	2	...	...	86 k	29 b	0	0	2	6	0	0	*Frost.
2	0	0	2	...	...	14"	6"	...	...	2	...	...	...	*Very dropsical.
5	4	0	1"	129 k	85 k	26 b	...	5	1	1	6	7	4	

The mode of inoculation was one which I learned from my friend Dr. Sanderson: to take a very small trocar fitted with a piston longer than the trocar; push up the sharp end of the trocar a small piece of the material (and I have never introduced a piece larger than a canary-seed or hemp-seed); introduce the trocar under the skin, and push the piston down till the material is delivered under the skin, and retained there. When fluid is used, I have used a small injecting-syringe with rather a large nozzle. I have never, under the latter circumstances, used more than two or three drops. The site of inoculation has almost invariably been the nape of the neck.

I may add that I placed the animals under healthy hygienic conditions in a large wooden shed, looking to the south, thickly protected with felt from the frost, provided with a window, and with access to a paved and sloping yard, exposed during the whole day to the sun. The house was divided into two parts—one containing the animals that had been inoculated, and one for the others. They were abundantly supplied with food, and their conditions have been, I think, as healthy as possible.

I will now proceed to describe in detail the anatomical effects which have followed the introduction of the materials in this manner. I must, however, state at the outset, that the appearances now to be mentioned as having been observed either locally or in internal organs, apply to all cases in which I have succeeded in producing what I consider tubercle, whatever be the material employed; for except in those cases where the inoculations were ineffectual in producing changes in internal organs, no differences could be observed either by the naked eye, or by microscopic observation, in the effects which followed.

To commence with the local effects, there are two which are rare. One of these is, that a slough is produced in the neighbourhood, which sometimes proves speedily fatal, with no further disease of organs than an inflammatory enlargement of the neighbouring lymphatics, totally different in its character from what I consider as a tuberculous change. The lymphatics are then swollen, soft, and more opaque than natural, but not cheesy, indurated, or suppurating. The other, which also is rare, is local suppuration. Now, in the guinea-pig, local suppuration—that is, the formation of diffused or circumscribed abscesses—is, comparatively with the rabbit, extremely rare. In the rabbit, it appears to be more common. The most common result is the formation of masses of cheesy matter, which are dry and friable, sometimes offensive when ulceration has taken place externally through the skin, but hardly ever presenting a soft puriform fluid. Under the microscope, the material so formed has all the usual characters of cheesy matter, showing dried-up *débris* of fattily degenerated material, rather than true puriform cells. The most common effect by far is the production of this cheesy matter under the skin. It is usually encapsuled; and sometimes there are two or three such masses, distinct from one another, or they may be more numerous, and may, together with the granulations next to be described, be found at considerable distances from the site of the inoculation. Thus in one instance they extended two-thirds of the length of the spinal column, and caused death through paralysis consequent on their pressure producing erosion of the bodies of two vertebræ, and subsequent softening of the spinal cord. In another case I found a rib similarly injured, but in neither of these instances was there any distinct tubercular change in the bone. The capsule of these masses is firm, semi-transparent, almost lardaceous in appearance; it passes insensibly into the surrounding tissue.

In addition, however, to these cheesy changes, another set of appearances are seen around the seat of injury. These are small granulations, varying in size from a poppy-seed, or even smaller, to that of a hemp-seed, which are irregularly scattered for a variable area in the subcutaneous tissue. They are sometimes semi-transparent throughout; sometimes yellow and opaque throughout; sometimes yellow in the centre, with a semi-transparent margin. Larger masses, of the size of a pea, are sometimes also found, apparently presenting intermediate gradations between these and the larger cheesy masses: for they are cheesy in some parts, and indurated in others; and they present a striking resemblance to the naked eye, and also under the microscope, to the changed lymphatic glands. These larger granulations are most commonly found near the seat of injury. The smaller granulations are often scattered on the outside of the capsules of the larger cheesy masses; but they may extend to great distances beyond this, sometimes in lines and rows, sometimes in scattered circles, and in little groups. They often reach up to the neighbouring lymphatic glands, and are grouped in masses of variable size around these (Plate I. fig. 6).

In addition to these, cords of induration are seen extending, at variable depths and for variable distances, through the subcutaneous and muscular tissues. The cords are firm and semi-transparent, like gristle; but their central portions are often cheesy. Sometimes they form varicose dilatations, reaching up to the lymphatic glands. Sometimes cords extend through the lymphatic glands into the surrounding tissues: whether from or to the gland, it is difficult to say (Plate I. fig. 7).

The cords are not constant, nor are the granulations, but the frequency with which they occur is very considerable. The granulations have not been very largely noticed, except by M. Villemin, who describes them as frequent. I believe the reason for this has been, that, to a great extent, they are seated deeply in the subcutaneous tissue immediately under the hairy skin; and, unless they are looked for carefully, and all the superjacent tissue removed, they may easily escape observation.

When these little granulations are examined under the microscope, they are found to present the following appearances. In the centre, they appear to present, for the most part, nothing but a mass of nuclei, or rather, I should say, speaking my own conviction, of cells imbedded in a homogeneous tissue, of which the nuclei alone are visible; or it may be stated that, in this part, they consist of nuclei imbedded in a homogeneous tissue. Further from the centre, and where their structure is less dense, these granulations may be seen to consist of round, or occasionally of fusiform cells, imbedded in the meshes of a fibrillated tissue, which forms bands or trabeculæ, between which the cells lie. The bands are of variable thickness, and are subdivided into narrower bundles. A series of spaces are thus formed, which are filled with rounded nucleated cells, but, in many parts, single cells having this character are surrounded by this fibrous network (see Plate III. fig. 8). A structure is thus produced, having the strongest resemblance to the elementary composition of a lymphatic gland, or to the cytogenic tissue of it.[1] The development in this manner from the elements of the connective tissue of a structure having so strong a resemblance to a lymphatic gland in miniature, is a very interesting fact among this series of pathological phenomena.

His

Towards the circumference of these masses, and in the adjacent tissue, where the reticular arrangement is less distinctly marked, fusiform cells, and oval cells, often with double nuclei, may be seen in various stages of multiplication and division.

The ordinary size of the cells constituting the granulations is from 1-2000th to 1-3000th of an inch in diameter. Some larger cells may reach a size of 1-1000th by 1-800th of an inch. The nuclei are very uniform in size—from 1-3000th to 1-3200th—and their outline is particularly sharp and well defined. Their contents, too, are refractive and glistening. Sometimes one or more nucleoli may be seen in their interior. The larger cells are sometimes finely granular. They present a certain resemblance to epithelial cells—a resemblance which has also been noticed by Professor Lebert.

Throughout the granulations many of the cells, and also of the nuclei, are seen in various stages of fatty degeneration. The fat is sometimes in

[1] Untersuchungen über den Bau der Lymphdrüsen, 1860.

visible drops; sometimes in a finely molecular form. The change is most common towards the centre of the granulations, but is also seen in cells irregularly scattered through the adjacent tissue.

There may also be sometimes seen in these masses, and in their neighbourhood, extraordinary strings and rows of cells and nuclei. Sometimes cells apparently, sometimes only nuclei, are visible. The nuclei correspond with those last described. The cells, when visible, are of the type of the smaller round cells; the row or string is contained, apparently, within some limitary membrane. The size of these rows is nearly twice or three times that of a capillary. They have no fibrous investment like a vein; no muscular coat like an artery. They suggest to me the idea of a lymphatic tube, swollen either by growth within its interior, or by cells carried into it and obstructing its interior. My own conviction is, that they are lymphatics, filled with a growth of cells (Plate III. fig. 6).

When we consider the omentum, I shall have to describe to you structures to which these granulations present, especially in their naked-eye characters, the most striking resemblance. The granulations under the skin are, however, denser; and I have not found them situated in the sheaths of the vessels, as they commonly are in the omentum.

The next change of importance is in the lymphatic glands in the neighbourhood of the injury—the axillary and the subscapular. They are commonly enlarged to twice or three times their natural size. The changes affecting them are also found in those at a considerable distance, as in the submaxillary, the infraclavicular, and the substernal glands, and in those situated in the course of the trachea. Their apparent number is also increased—that is to say, lymphatic glands of the size of a horse-bean, or even of a kidney-bean, are to be seen where none are visible naturally. On section, they are indurated, semi-transparent, cartilaginous-looking, and showing very little distinction between the cortical and medullary substances. Through them are scattered specks, lines, or streaks of cheesy degeneration, of an opaque yellow colour, which are sometimes agglomerated into areas of the size of a pea (Plate I. fig. 5). These sometimes soften into a creamy, diffluent matter, which presents the same granular material and molecular *débris* which are found in the cheesy masses surrounding the wound. Occasionally, most of the lymphatic glands throughout the body may be found to have undergone similar changes. These may be seen in the lumbar, sacral, and inguinal lymphatics, without any corresponding alterations being found in adjacent parts sufficient to account for the diseased state of the glands.

c

The chief changes which can be observed by the microscope in the lymphatic glands are, that there are large agglomerations of cells in groups in the cortical substance, apparently cramming up the loculi with structures precisely resembling the ordinary lymphatic cells, but much more densely packed; and these in some places have undergone fatty degeneration. At other parts, there are large tracts which have undergone a fibroid change. Among the meshes of fibres in these parts are cells which are not distinguishable from the ordinary lymphatic cells; but both in these, and also in the parts where the cells are denser, larger cells may be found, of 1-1000th to 1-2000th of an inch in diameter, containing one or more nuclei, and often undergoing a fine molecular degeneration.

The next organs in which changes are most frequently found are the lungs (Plate I. fig. 1). The chief state I have to describe consists in these organs being permeated more or less thickly by scattered granulations. These vary in size from a millet-seed to a hemp-seed. Some are very small, minute specks, scarcely visible by the naked eye, and all gradations can be found between the smaller and the larger. They are generally scattered, but are sometimes confluent in groups. Sometimes larger groups are confluent, especially at the edges of the lobes; but even in these there is more or less evidence that they have been originally composed of distinct granulations. The granulations do not project much from the cut surface, and they blend more or less intimately with the surrounding pulmonary tissue. They are all marked by a peculiar, semi-transparent, hyaline, cartilaginous-looking margin, and a cheesy centre. Some of the smallest may be found semi-transparent throughout. They are firm, and they tear with difficulty from the surrounding pulmonary tissue. As a rule, the centre is only cheesy, not soft; but, in some marked instances, where they have attained the size of a small pea, I have observed them with a distinct softened centre, which, when evacuated, left a distinct cavity. Lebert,[1] in one of his observations, found them breaking down into cavities of considerable size. They are more common on the pleural surface than deep in the lung, but are distributed pretty equally, not preponderating in one lobe more than in another. Sometimes, when there is a group of these granulations clustered together, an appearance is presented, which is also occasionally seen in the tubercle of the human lung, of a fibrous network running between the granules, as if the intervening tissue was becoming fibrous. In addition to this, there is sometimes a general induration of the lung-tissue, affecting, in a variable degree, the whole organ. Signs of pneumonia, and of general infiltration independent of the granulations, is exceed-

[1] Virchow's Archiv, xli.

ingly rare. Once or twice I have seen it in the rabbit; never distinctly in the guinea-pig.

The microscopic examination of these growths in the lungs presents the following features. In the first place, there are three main points in which they appear to originate. One of these is around the bronchi, another is around the blood-vessels, and another is in the tissue of the lungs, where no particular connexion can be seen with either bronchi or vessels. Around the bronchi (Plate II. fig. 1, A, B), they seem to extend from little masses of a lymphatic character, which normally exist in the bronchial sheath, and which are abundant in the guinea-pig, and are also stated by Kölliker to exist in the human lung. These granulations in the bronchi consist of masses of cells growing in their sheaths, 1-2500th to 1-3000th of an inch in diameter, mostly round, but sometimes, when densely packed, showing nothing but nuclei. These cells extend into the surrounding pulmonary tissue, and produce effects which I will dwell on presently.

The other place in which they commonly originate is in the perivascular sheaths of the pulmonary arteries (Plate II. figs. 2, 3). Here the growth appears to be nothing more than an accumulation of the cells lining the perivascular canal (fig. 2). The growth may extend for a considerable distance in length along both the peribronchial and perivascular sheaths; and from both these sources of origin a rapid extension ensues into the surrounding walls of the alveoli and smaller bronchi. A thickening of these is thus produced, and apparently by a double mode of growth—by a rapid development of fusiform cells at the margins, clusters of which are seen passing among the capillaries, and by an increase of rounder cells, which are seen nearer the centre of the new formation. Coincidently with this growth, a change of great importance occurs in the capillaries of the lungs (Plate II. figs. 5, 6, 8). Their nuclei enlarge, and the vessels, otherwise apparently unchanged, contain no more blood; that is to say, no injection will pass into them; and yet their outline is still marked, even in the neighbourhood of these tubercular masses, by lines of nuclei traversing the base of the air-vesicles. This obstruction of the capillaries takes place through very much wider areas than the space apparently occupied by the tubercle. So also with the thickening of the walls of the alveoli. Around the grey granulations, and for a space of three or four times their area, there is a circumference of thickening affecting both the walls of the alveoli and of the smaller bronchi. This appearance is more distinct in the guinea-pig than in many specimens of tubercle in the human lung; but it can be distinctly seen sometimes in the latter, and is a fact of great pathological importance, explaining the increased density and loss of elasticity of the lung which occurs in the early stages of tuberculosis, and which cannot be satisfactorily accounted for by the mere presence of the grey granulations (Plate II. fig. 4).

The bronchial glands are affected exactly like the axillary, and a description of the one applies precisely to the other, except that the cheesy changes are rather less marked, and spots of softening are less common in the bronchial glands.

The next organ which presents marked changes is the liver (Plate I. fig. 4). It is greatly increased in size, and is heavier than natural. In it are seen two or three sets of changes, which, however, I think, all belong to one type. The chief of these is the appearance of a tissue of a glistening semi-transparent character, which occupies the substance of the organ to a variable extent. Sometimes it only occurs in little specks, of the size of a pin's head or a poppy-seed; sometimes it is in larger masses; sometimes it runs in lines, mapping out the acini very distinctly; sometimes it is found in irregular masses in which there is no trace of liver-structure visible to the naked eye; or the conditions before described may be reversed, and remains of natural-looking liver-tissue may be seen running in lines irregularly through the semi-transparent tissue.

In the first of these cases, the disease might almost suggest the idea of cirrhosis; in the second, that of an albuminoid liver; but even to the naked eye it differs from the cirrhosed liver in being crisp and friable and destitute of all toughness, and from both it and the albuminoid liver in the fracture being finely granular. This tissue never stains with iodine; and it almost invariably has an appearance which shows, more or less distinctly, that it was originally composed of fine granulations. Moreover, under the microscope the change is seen to be of a kind absolutely different from that which occurs in the larda-ceous liver, inasmuch as here there is no sign of the waxy change in the secreting cells which characterises the lardaceous disease. In other cases, the granulations are not confluent, but are scattered thickly through the liver-tissue. They do not project notably from the surface, and the peritoneum is not, as a rule, thickened.

In some places, there are seen, irregularly scattered in the liver-tissue, spots of a dead opaque white, sometimes surrounded by a zone of injection. Their section is finely granular, not smooth like those next to be described, and they appear to result from a fatty degeneration of the acini.

There are also seen, both in the larger tracts of semi-transparent tissue, and also when this change only occurs in the form of granulations, spots of cheesy change. These may be either seen occupying the area only of individual granulations, or they may extend through considerable spaces. They generally, however, remain firm and hard. In some cases—though the appearance is rare—they may form masses of the size of a hazel-nut, with a diffluent creamy centre.

When sections of the liver are examined with the microscope (Plate III. figs. 4, 5), the semi-transparent portions, whether in granules or in confluent masses, are found to consist of a new growth situated in the interlobular tissue, or in what is called the capsule of Glisson, between the acini. The tissue thus produced consists of cells uniformly imbedded in a delicate fibrous network, in the meshes of which they lie. In size, the cells correspond to those found in the new growths in other situations; the ordinary size being from 1-3000th to 1,3500th of an inch in diameter. Some may reach the size of the 1-2000th to 1-2500th of an inch. The nuclei average about the 1-3600th of an inch. These are strongly refracting, and have a distinct margin. The contents of the cells are nebulous, and their outline is less distinct than that of the nuclei. In some of the larger cells, double nuclei may occasionally be seen. In some places, the cells are agglomerated into strings or rows, forming elongated masses resembling those described as seen in the granulations occurring under the skin, and to which the same interpretation is probably applicable.

The amount of this tissue between the acini varies. Sometimes these are occluded over large spaces, or outlines only of the acini pressed aside may remain mapping out the masses of new formation. The sections under the microscope show that these latter always tend to assume a globular form, and that at their margins the cells of new growth, which are strikingly different from the liver-cells, pass up in lines between the network of cells of which the acini of the liver are composed (d d d, fig. 4). The liver-cells appear to atrophy in consequence of the progress of this growth, but they undergo no further changes. There is no multiplication of their nuclei, as if they participated in the growth, and they undergo no special granular change. They become pale and nebulous, with a fine molecular appearance; and here and there some fatty change when they are pressed upon by the growth. Both the bile-ducts and the proper gland-tissue appear to be otherwise unaffected and healthy. Little specks of fine granular degeneration may be seen scattered through the masses of new growth, which in some spots pass into a more complete molecular *débris*.

The spleen (Plate I. figs. 2, 3) also enlarges very much; sometimes to two, three, or four times its natural size. I have measured it two inches and a quarter long, one and a quarter broad, and half an inch thick. It also is greatly increased in weight. This organ also presents various appearances. A very common one is, that small whitish spots are scattered thickly through the whole tissue. These appear of about twice the size of the normal Malpighian bodies. Another consists of agglomerations of these rounded bodies into masses, which have then a racemose appearance.

Other masses are seen more opaque, but having the same structure of agglomerated granules. The spots of smaller size have generally a certain transparency, but less than is observed in the liver or in the lung. Sometimes large tracts of the spleen are converted into a semi-transparent brittle tissue resembling that of the liver or of the lymphatic glands, and showing also an appearance as if originally composed of fine granulations.

Various forms of cheesy change are also found. These in some places occupy about the same area as the smaller granules, and are seen as scattered, yellow, dry, firm spots, brittle, and crumbling under pressure into a granular mass. The larger compound granulations may similarly become cheesy and opaque. Cheesy spots, of variable size, may also be seen scattered through the semi-transparent tissue. In rarer cases, large irregular tracts may become cheesy, but still maintaining the same firm, dry, friable, and opaque character. In yet rarer instances, large portions of the spleen are found changed, like those of the liver, into masses of the size of a hazel-nut (which is the largest I have seen), and composed of a diffluent softened amorphous material; but, in such cases, semi-transparent granulations and cheesy granulations are seen in other parts of the organ.

The large cheesy masses are not very distinctly separated from the surrounding tissue, and are not surrounded by any zone of injection. They differ also considerably in the character of the softened material and in the absence of abruptness in their limitation from anything known as pyæmic spots. Under the microscope, most of the granulations seem to be composed of enlarged Malpighian bodies undergoing various stages of fatty degeneration. The whole tissue is also infiltrated with cells precisely resembling the lymphatic cells of the spleen. Together with this increase of cells, there may be found, in the more transparent areas, fibrous changes analogous to those described in the lymphatic glands.

The changes in the intestines mostly affect the small intestines and the cæcum. In the former, the patches of Peyer are enlarged, and they are prominent externally. The peritoneal surface is injected, and sometimes thickened, and through it numerous opaque white spots can be seen. Internally, also, the patches are enlarged and prominent, and the mucous membrane around is injected and opaque. Milky-white spots are seen scattered in the substance of the patches. Ulceration also occurs in specks in them, but it seldom affects the whole patch.

In the cæcum, the patches are also much enlarged; but the single follicles are less distinct than in the small intestine. Ulceration of the whole or of portions of the patch are not uncommon; and, when present, they have often a cheesy floor, in which are numerous small granulations, like those seen in the floor of a tuberculous intestinal ulcer in man. Thickening of the margin is less marked; but in explanation of this, it may be stated that the mucous membrane of this portion of the intestine is much thinner in the guinea-pig than in man. With this exception, they present all the characters of the tuberculous as contrasted with the typhoid ulcer of the intestine. The solitary glands of the small intestine are occasionally enlarged; sometimes also they are cheesy.

Microscopic examination of Peyer's patches in the stage of enlargement shows cells crowded in the follicles in a manner similar to that observed in the lymphatic glands and in the spleen, and in many parts undergoing fatty degeneration. Once I have seen a bloody slough adhering to the ulcerated surface, and once punctiform extravasations in other parts of the mucous membrane of the intestines. With the last exception, I have seen no other signs of hæmorrhage. The ulcers always spring from Peyer's patches or from the solitary glands, and not from other parts of the intestine. The stomach has always been unaffected.

The mesenteric glands are almost invariably affected, even when distinct changes are not discoverable in the intestine itself. One gland shows this most prominently; viz., the large one situated in the part of the mesentery attached to the ileo-cæcal valve. The large glands in the hilus of the liver and the gastro-hepatic omentum are also very commonly affected. Their changes are similar to those of the axillary glands.

In the peritoneal cavity, dropsy is very common. Lebert attributes it to the obstruction of the circulation through the liver, and I think with reason. I have once seen hæmorrhage into the cavity. I could not discover its source, but all the branches of the portal vein were loaded with blood. The chief change, however, to which I wish to direct your attention in this part occurs in the omentum. Here, scattered thickly through the tissue, are found granules varying in size from a pin's point to a poppy-seed, seldom much larger. They precisely resemble the grey granulations found in tuberculous peritonitis in man, except that I have only once found any sign of inflammatory action attending the process. This was in a case of general tuberculous peritonitis following abortion.

Under the microscope (Plate III. figs. 1, 2, 3), the granulations are found scattered in the course of the vessels, and they then originate in the perivascular sheath. They are also, however, to be found in parts of the tissue where no vessels whatever are to be seen, and here they appear to have their starting point from masses of nuclei or cells agglomerated among the meshes of the omentum, which appear normally to occur in scattered points of this tissue. From both sources, the growth extends irregularly to the surrounding tissue; and whatever their origin, whether from the sheaths of the vessels, or independently of these, the only further change discoverable consists in a dense agglomeration, most commonly of the nuclei, and sometimes of the cells, in the meshes of the network of which the omentum is composed. With this exception, I have never been able to trace any distinctive characters which separate the histological characters of tubercle in the omentum from that of the healthy surrounding tissues. I have never seen anything like an epithelium covering the omentum, as distinct from any nucleated tissue beneath. The only appearance which we can see in the normal omentum is that of cells (the outline of which is often indistinct, and of which the nuclei alone may be visible) imbedded in the meshes of a fibrillated tissue—a structure which strongly resembles that which ordinarily is called "lymphatic tissue." The cells, both in the normal and also in the tubercular omentum, vary somewhat in size; but their average is between 1-2400th and 1-3200th of an inch in diameter.

To present you briefly with a summary of the number of times in which these appearances have been found. Out of the 117 cases, three internal organs have been affected fifty-eight times; in six, less than three. Out of these sixty-four cases, local granulations at the site of injury have been stated to be present forty-one times. They were absent in all the doubtful cases. They were absent in the case of general tubercle following the insertion of a cotton-thread, though the fibrous tissue in which this was imbedded presented some indistinct traces of their appearance. They were absent, also, in three out of four cases, where tubercle followed the insertion of a cotton-thread saturated with vaccine-lymph. They were never present when no certain internal effect was produced, except in one doubtful case in a rabbit.

The local lymphatic glands are noted to have been affected with induration and cheesy spots in fifty-four cases. In one or two of the doubtful cases, there was induration; in two others, enlargement, but no other change. In three others of the doubtful cases, they were distinctly affected. They are only stated to have been absent in four cases of general tuberculosis; and I should almost question the accuracy of my own observations, whether I may not have overlooked one or two enlarged glands in these instances.

The lungs have been affected in fifty-nine out of the sixty-four. In one, they were unchanged, when the spleen alone was affected; in one, when the liver, spleen, intestines, and mesentery were affected; in one, when the liver and spleen suffered; in another case, the affection was doubtful in its nature.

The bronchial glands escaped only eight times out of the sixty-four; twice only when there was any considerable lung-affection present.

The spleen has been affected fifty-nine times, only escaping five times out of the whole group.

The liver has been affected fifty-one times. In nine, the character of the affection was considered doubtful, consisting only of cheesy spots, which in some cases appeared to be due to psorosperms, and where no other distinct affection could be discovered. In four, it was altogether absent.

Of the omentum my notes are less perfect. It was affected twenty-five times. The condition was not stated in twelve cases.

The intestines were affected in twelve cases; not examined in twenty-two. In the rest, though examined, no affection was found.

The mesenteric glands were affected forty-five times out of the sixty-four.

In two cases granulations, having all the character of grey tubercle, were found sparsely scattered in the kidney. Microscopic examination showed that these growths corresponded with the grey tubercular granulations found in this organ in man. They consisted of small cells and nuclei imbedded in a fibrous matrix, and they were situated in the intertubular tissues.

In two cases the uterus was found in a state of catarrhal inflammation, and in one of these its interior was found filled with a cheesy matter, much resembling in this respect a drawing of Carswell's of tubercle of the uterus.[1] Microscopic examination showed small granulations having the ordinary tubercular character in the submucous and intertubular tissues. The glands presented the ordinary appearances of catarrhal change, having their epithelium swollen and granular. In the case last mentioned the retro-peritoneal and lumbar lymphatics were enlarged, indurated, and cheesy, thus presenting appearances similar to those before described as occurring in other parts.

In one or two cases opaque spots were found in the cornea. In one of these, which Mr. Vernon had the kindness to examine for me, there were seen cells of new formation imbedded among the corneal fibres and under-going fatty degeneration. Neither in this case nor in many others which

---

[1] *Loc. cit.* Tubercle, Pl. II. figs. 1, 2.

D

I examined could I find any granulations in the choroid, and the meningeal arteries in all cases examined appeared to be free from anything like tubercular growths.

To contrast these results with twelve cases of inoculation in rabbits, I inoculated five with pus without producing any general affection; but in nearly all these there was extensive local suppuration under the skin, producing in many cases large abscesses, which when old were filled with diffluent curdy matter, but were not associated with the granulations described in the guinea-pig. Microscopic examination of these showed that they consisted of pus alone, and they presented none of the appearances described as characterising the subcutaneous granulations.

Inoculations with tubercle also on the rabbits, seven in number, only produced distinct effects in two: and these were mixed with pneumonia and with other internal inflammations, which had some appearance of being of pyæmic origin, and in three cases with a laryngo-tracheitis extending to the lungs, and having a strong resemblance to diphtheria in the adult (Pl. I. figs. 8, 9, 10). I mention these to show that the tubercular affection in the guinea-pig is much more easily produced than in the rabbit.

The most important question, however, in relation to these experiments is that of the correspondence of these growths with what is ordinarily considered as tubercle; and, in discussing this, I will take as the type of tubercle that pathological product concerning the nature of which there is the least difference of opinion—viz., the grey granulation. There are, however, two or three leading points which I should wish to lay before you with some emphasis. One of these includes the mode of distribution, the multiplicity of the affection, the organs affected, and the individual parts of the organs so affected, which strikingly correspond to the chosen seats of tuberculous affection in man. There is one exception in the guinea-pig—viz., the absence of meningeal affection; but I do not think that this can weigh very materially against the cumulative force of the argument derived from the other viscera, especially as tubercular changes in the meninges do not form a necessary accompaniment even of the process of general and acute tuberculosis in man. It is possible also that the non-implication of the meningeal arteries in the guinea-pig may depend on minute points in their anatomy with which we are not yet familiar.

I would also call your attention to the fact that, in all these organs, the main characteristic of the affection is the production of a series of growths resembling the lymphatic tissues, and probably having their starting-point in structures of this nature. Now I think that, if there is one definition of tubercle which is more tenable than another, it is that of Virchow, who has pointed out the analogy between tubercle and the lymphatic structures ; viz., that they both consist of round cells, of many of which the nuclei alone may be visible, imbedded in a network either of fibres or of a more transparent but semi-cartilaginous-looking homogeneous tissue ; in which, however, distinct fibrillation may be absent.

Another point is, that these growths are agglomerated into little masses, like the follicles of lymphatic glands, or the Malpighian bodies of the spleen, or the solitary glands of the intestines ; that, wherever they occur, they tend to form distinct granules, but that at their margins they blend insensibly with the surrounding tissue ; that they extend in connective tissue by a change in the mode of growth of this tissue, through which cells are produced resembling rather those of a lymphatic gland than of the connective tissue itself; that these cells have no specific character, in this resembling those of tubercle ; that they pass into an early decay, which is not sloughing, but molecular death ; and that this degenerative change appears to be caused partly by their mutual pressure and by the density of the growth, and partly by the occlusion of vessels. I think that we know of no other growth except tubercle of which all these attributes can be predicated, and certainly none other which occludes the vessels of the part to the extent which is effected by these formations in the guinea-pig.

The identity of the morphological structures of these growths in the lung with those of human tubercle may, I think, be affirmed with some distinctness. One fact may also, I think, be very distinctly stated—that we are not dealing with any "lobular pneumonias," or with any "catarrhal pneumonia," or with any infiltration of the interior of the alveoli (which are the very last parts that are occluded). The growth proceeds by a progressive thickening of the walls of the alveoli, which finally closes the air-vesicles. Even in the densest masses, when examined with a binocular, a certain transparency can be distinctly seen in some places, showing that the central portions of the alveoli are still hollow. There is very little participation of the epithelium in the growth. The epithelium seems, if I may say so, to disappear in the progress of tubercle, and to give place to cells of another and different character. I think (if I may digress for a moment) that we have been dealing with the question of the epithelium of the lung in too specific a sense ; and that in the alveoli, it is only a derivative of the connective tissue, like the cells which I have described in the normal omentum, and having but little resemblance to the appearance of an epithelium in a mucous membrane, which rests on a nucleated subjacent tissue. In the alveoli of the lung, no subjacent nucleated tissue can be seen, and the cells lining them appear to be the sole representatives of cell-forms (with the exception of the nuclei of the capillaries and of some fusiform nuclei), in the delicate membrane of which the wall of the alveoli is composed. Under certain morbid processes, especially of the catarrhal type, they enlarge and desquamate; and we then call them epithelial cells which have separated. At other times, however, the structure no longer produces epithelial cells, but cells which are the direct derivatives of connective tissue. A few large cells like epithelium are occasionally produced; but these latter do not, as a rule, form a large proportion of the grey granulation in the lung. The epithelium passes away, or, at any rate, does not maintain its type in the process of the new growth. There is no filling of the alveoli—at least, not to any proportionate extent—with epithelial forms; and thus far we have not been dealing with anything like ordinary pneumonia, and the process is absolutely dissimilar from that of pyæmia.

It appears also that, in the lung, the point of departure of the growth is in no essential respect from the vascular apparatus; nor is this implicated in the same manner as it is in most inflammatory processes. The point of departure, even when it is connected with the vessels, is in a tissue external to the vessel; and the occlusion of the circulation which occurs round the growth is either through pressure, or by the implication of the walls of the capillaries in the growth; and it passes backwards from them to the smaller arteries, and not in the converse direction.

I think that there can be very little question but that the affection in the lymphatic glands is one so closely akin to the tubercular change, that I need not dwell upon it further.

The organ which might excite most doubt is the liver, because, in man, tubercular growths here very seldom proceed to the same extent as is observed in the guinea-pig. But really, in its essential characters, this growth appears to me to be identical with that of tubercle in the human liver. Their seat is the same; their relation to the acini is the same; and in both we see the same immunity of the proper liver-structures from all participation in the process of the new growth; and I think I may say that we know of no other disease except tubercle to which the description I have given of the growth in the liver can apply. The only one producing at all analogous effects in the liver is leucocythæmia, and to that I will allude in a few moments.

As regards again the omentum, any one who will compare for a moment one of these growths with specimens of tubercular granulations from the human omentum, or from that of the monkey, can have, I think, but very little doubt remaining as to their perfect histological identity. They are in both cases perivascular, and also extravascular, and they both equally pass gradually into cheesy decay.

I must confess that, sceptical as every one must naturally at first feel on this subject, the cumulative force of the evidence in favour of the tubercular nature of these growths appears to me to be irresistible. We are either dealing with tubercle, or we have before us a new and hitherto unknown constitutional disease of the rodentia, consisting of growths which, in their naked-eye appearances and histological characters, correspond with all the essential features of tubercle in man; which occur not only in the organs which are the chosen seats of tubercle in man, but also in the same *parts* of those organs; which have the same vital characters, and the same early degenerative cheesy changes—not suppuration nor acute softening—and with no marked characters sufficient to distinguish them from tubercle. It would appear to me that, according to all our ordinarily received rules of pathological definition and classification, this disease must be considered tubercular; for the analogy is not one of mere histological refinements, but also of seat of selection, mode of growth, and vital characters: and, therefore, extraordinary as its mode of production may appear, we are not on that account justified in excluding it from the pathological category to which it appears properly to belong. The only known disease which possesses the least affinity with it is leucocythæmia; but leucocythæmic growths are softer, whiter, and more milky or medullary in appearance than these granulations, and they have not the same tendency to degenerative changes, and they are, as an almost invariable rule, attended with an increase of the white corpuscles of the blood, which these growths are not. Tubercle also is similarly unattended, though such an increase might be expected from the implication of the lymphatic apparatus.[1]

I do not think that the disease can be considered as by any means identical with farcy, though the local changes may bear some resemblance to farcy changes; but every subsequent change of farcy is one of a suppurative kind in internal organs.

After the recent specimens which I have submitted to you, I do not think that any one can maintain that these changes can be included under anything ordinarily understood by the term pyæmia.

The great difficulty in the present day in identifying any disease with tubercle depends, I think, on defective, and, as it seems to me, somewhat

[1] Billroth, Beiträge, p. 147.

arbitrary definitions of tubercle. I think, if I may pass such a criticism without presumption, that some modern pathologists are rather disposed too strictly to limit the application of the term tubercle. Even Virchow's careful observation, to which we owe so much—that *all* cheesy matter is not tubercular—has been since pushed to an almost dangerous extreme; so that some are even disposed to doubt the tubercular character of any cheesy matter; whereas, on the contrary, tubercle is among the most frequent, though not the sole cause of such products in the body. Indeed, these limitations of tubercle have proceeded so far that, if the exclusion of the different forms from the category of tubercle proposed by various pathologists were simultaneously carried out, tubercle would, not unfortunately, cease to exist, but would certainly no longer have any place in our nosologies; for nearly every pathological product hitherto ranked under this title, from the grey granulation to the yellow granulation and the cheesy infiltration, is by some authority or other excluded from the category of tubercle. I would not for a moment say that I think that no limitations to the class of tubercle are desirable; but I think that we are at present in danger of carrying these to an extreme degree. My impression is, that tubercle is a product which undergoes transformation in various directions; but that, though it varies somewhat in appearance, especially in the lung, according to the rapidity of its development or the greater or less amount of implication of epithelial structures in the process of its growth, and though it may, at one stage of its growth or transformation, present appearances different from those observed at another, yet that these stages and varieties of appearance are not the less tubercle.

There is, however, another point to which I desire especially to call your attention, in relation to the question of the identity of the disease in the guinea-pig and in man. This question appears to me to rest on a broader basis of analogy than that of the correspondence in every line and shade of description between the statements of individual observers, as regards either the histological characters of these growths in the rodentia or those of tubercle in man. Looking at the general history of histological research, I cannot for a moment flatter myself that I have exhausted all the points in the minute anatomy of these new formations; though I think that I have, so to speak, been able to lay before you the leading features of their ultimate structure. But what I wish to insist on is the *general or constitutional* character of the affection, which coupled with the general analogy of the new growths with tubercle, I think, almost absolutely proves their tubercular nature. It is not a question of the lung *alone*, or of the liver *alone*, or of the lymphatic glands, or the spleen, or the omentum or intestines considered singly. It is a question of a *general* disease, producing in all these organs growths which, if they occurred in man, would by ordinary observation be considered tubercular; and as no other disease is known except tubercle which produces these effects, I therefore think that, considering the identity of these with all the most important features of this disease, we cannot but admit the tubercular character of the disease thus artificially produced.

I would now desire briefly to call your attention to the mode of the production of these growths. It is not my intention to propose any distinctly new theory, for the subject has received so many theories already, that it would require a considerable exercise of mental ingenuity to found a new one. Still, however, there are, I think, certain legitimate conclusions which may be drawn from these experiments, if it be admitted that these growths are tubercle.

In the first place, M. Villemin's position that tubercle is a specific disease, producible by tubercle alone, cannot, I think, be held to be true; nor can the method of inoculation be used as a test of the tubercular character of any pathological product, for the four guinea-pigs in whom the vaccine lymph was inoculated, and those inoculated with putrid muscle, and even one beneath whose skin I simply inserted a piece of cotton-thread, and also one of the four in which, following Dr. Sanderson's example, I inserted a seton, presented as intense and typical specimens of the disease, as those on whom inoculation had been practised with the most typical grey granulations from the lungs or the meningeal vessels.

Secondly, the results of all experiments hitherto conducted, as far as they have yet been carried, appear to show that, for the production of the disease, septic matters in a certain state, introduced into or produced within the economy, are necessary. I say septic matters; for I am inclined to believe that the effect of the seton, or even of the cotton-thread introduced under the skin, is produced by these substances setting up an inflammatory action, the products of which have the same influence as the other unhealthy substances which the list contains. The failure in many cases of these inert matters to produce such effects, would appear to my mind to bear out this view, which is especially corroborated by the cases where setons were inserted. In three of these, the setons ulcerated out and the skin healed, and all these cases proved failures, showing that the infecting property probably depends on a certain kind of inflammatory action produced.

I may here, perhaps, be allowed to advert to the series of M. Lebert's experiments with matters directly introduced into the circulation. He injected mercury and charcoal into the vessels; and also quotes two experiments of

older date, in which, after repeated injection of pus into the veins, he found
what he considered tubercles in the lungs and liver. M. Lebert argues that
the effect of all these experiments was to produce tubercle; but I think that
a striking difference exists between the results of these experiments and the
effects of inoculation, and the difference I notice is this, that there is no
evidence to be found of a constitutional affection having been produced in any
of these cases whose details he has given at length. It is on the *constitutional*
affection, and the *multiple* implication of many organs, that I lay especial
stress in reasoning on the proof of the tubercular affection produced by
inoculation. Multiple affections did not exist in any of this class of M.
Lebert's experiments, with the exception that, in the two dogs into whose
veins pus was injected, some granulations, like tubercle, were found in the
lungs and liver; but the authenticity of the tuberculous nature of these was
not decisively established by microscopic proof. Perhaps, also, I may add
that, in the innumerable experiments of this nature that have been conducted
with a view to the illustration of the nature of pyæmia, the production of
tubercles—*i.e.*, of grey granulations—has never been noticed, or, if at all,
they have not been observed with a frequency sufficient to establish the
possibility of this method of their origin.

The experiments with charcoal and mercury appear to me to be equally
inconclusive. Out of ten experiments of this nature, in four only was there
a production of local granulations in the lungs, which originated in thickening
around the obstructed vessel, but sometimes extended beyond this; in one
only of these there were found a few granulations in adhesions of the pleura,
which did not contain mercury. In none, however, was there any implication
of other organs; and, in this respect, these cases present so marked a difference from the effects of inoculation, that I cannot but believe that in them
tuberculosis was not produced as a constitutional disease. Local thickening
around an obstructed vessel is not necessarily tubercle; and the same criticism
is, I think, applicable to the introduction of mercury, or of other irritating
substances into the bronchi. The effect has been a local inflammation; but
not either local tubercle or a general constitutional affection. Another
important difference to be noticed in these granulations is the fact recorded
by M. Lebert, that the vessels around them were not obstructed in the manner
observed in tubercle.

As to the manner in which septic substances act, there is room for a wide
diversity of opinion. One effect seems to be established, if the tubercular
character of the granulations under the skin be admitted—viz., that a local
irritant is capable of producing local tubercle; and the suspicion naturally

arises, whether this be not the starting-point of the whole process. If, however, the character of these granulations be doubted, still the affection of the lymphatic glands in their neighbourhood has so strong a resemblance to that which occurs in these structures, when secondary to tubercle in other parts, that it would appear either to confirm the tuberculous character of the granulations, or to carry the point of origin of the tubercular infection only one stage further forward.

One fact also deserves to be brought especially into prominence; viz., that the effects of infection are more certainly produced by the inoculation of tubercle than by that of other substances. The whole series of primary inoculations of this nature succeeded when the animals lived long enough to allow of the affection of their internal organs; and the series of twelve re-inoculations of tubercle artificially produced had the same results. It does not, however, appear that the mode in which tubercle of the latter class had originated produced any difference in the certainty of the effect of the re-inoculation. Tubercle developed by the inoculation of putrid muscle reproduced tubercle with the same certainty as that originating from the inoculation of the grey granulations.

Whether any direct effect can, in virtue of this superior infecting property, be ascribed to tubercle, must remain, I think, an open question; but it does not appear impossible that it may have in some way the power, when thus introduced, of directly propagating itself, similar to that observed in actual practice, when there is great probability that the multiplicity of the affection may be due to secondary infection of distant organs from that primarily implicated. Another point to be noticed with regard to tubercle—though this possibly may not prove very material—is, that it is capable of thus producing infection when inoculated in a perfectly fresh state from an animal just killed. It should also, however, be remembered, that the low forms of pneumonia, which were formerly classed with the tubercular products in the lung, appear to possess this infecting property in an equally high degree.

Whether the infection of the system is by chemical or by mechanical means, whether it is produced by fluids or solids, by cell-forms, "germinal matter," or amorphous material, must remain, I think, a matter of hypothesis, in which the minds of individual observers will incline to one view or the other, according to the bent of their special pathological theories, but which has not as yet received any absolutely satisfactory elucidation. An experiment of Dr. Waldenburg's, in relation to this question, is, however, deserving of mention. He has found that colouring matters introduced under the skin, together with the infecting material, are reproduced in the growths of new

E

formation; but even this scarcely appears to afford an absolute proof that the infection is communicated by solid particles, though the presumption would appear to be in favour of this hypothesis, since the absorption of colouring matter by new growths is only what is observed under many different circumstances in the animal economy.

I may, perhaps, be permitted to say, as will perhaps appear from the remarks that I have already made, that the morbid processes I have now been describing appear to me to be absolutely removed from the ordinary phenomena observed in the obstruction of vessels by emboli. I have already stated that the obstruction of the vessels is a consequence, and not a cause, of the growth, and the anatomical appearances observed differ widely from embolic phenomena. When a vessel is obstructed from within, there results a centre of anæmia and of early necrobiosis of the parts supplied by it, surrounded by a zone of injection; while here there is an area of growth, generally proceeding from certain limited centres, but surrounded by a zone, not of hyperæmia, but of irregularly extending anæmia.

In speaking, also, of septic absorption as a mode of production of tubercle, I think that we must admit that these changes are to be placed in a different category from the phenomena ordinarily classed under the term of pyæmia. There is not a sign of suppuration to be observed in these cases —at least, not in the internal organs; and so infrequent are the appearances of ordinary inflammation observed, that they may be almost said not to exist —at any rate, as an essential part of the process. M. Lebert speaks of the changes in the lung as a peri-bronchitis, or a peri-alveolitis, or a peri-arteritis; but all new growth is not inflammation. The inflammatory nature of tubercle, affirmed by Broussais and Cruveilhier, has been a ground of contention among pathologists since the writings of the former appeared; but at any rate, if the nature of the growth be inflammatory, it is an inflammation proceeding under special conditions, and with peculiar limitations to individual tissues and to special parts of individual organs. This latter condition, as well as its mode of origin, are further arguments against the theories either of embolism or pyæmia being applicable to this affection. Neither of these commence with local granulations, or, as a rule, with the peculiar implication of the local lymphatics here observed. Moreover, the parts of the organs affected further confirm this view. It would appear to be impossible to regard the peritoneal granulations as the results of either of these processes. Nor in the lungs, liver, spleen, or intestines, are either embolic or pyæmic conditions so specially confined to the lymphatic structures, as they are observed to be in these instances.

I must confess that the theory advanced by some recent pathologists, since Dr. Cohnheim's researches on the exit of the white corpuscles from the blood-vessels, that the growth of tubercle is a result of this process—a theory, be it remembered, originally to some extent advanced by Dr. Addison of Brighton—does not appear to me to be supported by observations on these growths. The complete obstruction of the vessels in their neighbourhood would militate against this view of their origin; and a considerable amount of evidence may be found in some parts of the growth, that their increase is due to the enlargement and subdivision of pre-existing cell-forms.

To sum up the conclusions which the facts appear to warrant regarding the origin of this process, I would say that the growths produced appear to originate from an abnormal nutritive activity of tissues either demonstrably lymphatic in their nature, or of those whose lymphatic character is a matter of probable inference; that this increased growth is caused by septic matter introduced into the system, and especially affecting these tissues by an irritative action; that the new growths thus produced conform in many particulars to the lymphatic type, but, growing under abnormal conditions, or with perverted activity, they speedily pass into degenerative decay. This action, however, once excited, tends to repeat itself in the economy, either by virtue of the results of such degeneration possessing properties similar to that of the material whose morbid influence first excited the diseased process, or through direct infecting properties of the new growth transmitted either by the blood to distant organs, or, as in the case of the lymphatic glands, by means of infection through the lymphatic vessels.

This theory of the effects of septic matter is not altogether new. It was originally propounded by Dittrich [1] as an explanation of the cause of tuberculisation, and again by Buhl; [2] the former of whom believed that the septic infection producing tuberculisation was due to the retrograde metamorphosis of any morbid pathological products; the latter attributing it only to retrograde changes in tubercle once produced. It finds, however, for the first time, its direct proof in these experiments. They also illustrate what clinical experience has recently led Professor Niemeyer to assert, that tubercle of the lung may be a secondary consequence of unabsorbed pneumonias.' They illustrate, also, what has been the result of frequent surgical experience, but was first, as far as I am aware, distinctly stated by Mr. Holmes, at the Children's

[1] Virchow, Krank. Gesch., i. 113.
[2] Zeitsch. Rat. Med., 1857.

' Klinische vorträge uber die Lungen-Schwind-sucht.

Hospital, that the removal of diseased bone[1] prevents the formation of tubercle in predisposed subjects.[2]

Although, as I have now stated, the evidence, as it at present stands, seems to show that these effects are most readily produced by septic agencies, it may yet appear not impossible that *ordinary* irritants may, under special circumstances, have the power of exciting locally that peculiar development of quasi-lymphatic tissues which, if not the essential character of tuberculosis, is yet its most constant and striking feature.[3] This theory, if true, would seem to explain many cases where catarrhal inflammation of the bronchi is followed by tubercular processes in the lungs, owing to the irritation being propagated to the lymphatic structures in these organs—events which have hitherto, in spite of very distinct evidence to this effect, been excluded by many writers from the causative influences of this disease, mainly through the views so largely held regarding its specific character.

There appears, however, to be another element necessary in these cases, and that is a constitutional state predisposing to such effects. This is well illustrated in the different susceptibility of the guinea-pig and the rabbit; the former being much more prone to be thus affected by these agencies than the latter. In fact, I think it probable that it is to this cause that the failure of M. Villemin's experiments with other materials than tubercle, and which were conducted upon rabbits, is to be attributed.

I think, however, it requires to be stated, that neither guinea-pigs nor rabbits are naturally by any means specially prone to tubercle. Zoologists and comparative anatomists, of whom I have asked the question, have not noticed this disease amongst them, except as a consequence of injuries. They have long been the subject of physiological experiment, but tubercle has not been noticed among their common diseases. M. Villemin's experiments also showed that only the inoculated rabbits thus suffered. I have kept large numbers of guinea-pigs during the last six months, and many rabbits; but not one has died tuberculous, except those inoculated. As a further test, I killed six non-inoculated guinea-pigs, taken without selection; and not one presented a trace

[1] *Lancet*, 1865, i. 60.

[2] My attention has, since the delivery of this lecture, been called to some earlier observations of Dr Von Troeltsch on this subject. In Virchow's Archiv, vol. xvii. 1859, pp. 61-77, he has related three cases of acute miliary tuberculosis consecutive to long-standing disease of the bones of the ear; and expressed his belief that the suppurative action thus induced was the cause of the tuber-

cular disease, which he considered was produced by the infection of the blood. See also Von Troeltsch, Anatomie des Ohres, 1861, p. 72; and Lehrbuch der Ohrenheilkunde, 3d edition, 1867, p. 357. Confirmatory cases have also been published by Schwartze, Archiv für Ohrenheilkunde, ii. 280.

[3] See Appendix, p. 31.

of the changes found in the inoculated animals. The failure also of nearly half of the experiments with inoculation, most of which animals were killed, and in which no trace of tubercle was found—viz., fifty-three animals out of one hundred and seventeen—proves strongly that tubercle is not naturally a common disease among them.

In what the predisposition may consist, must remain to some extent an open question. I admit that I am strongly inclined to regard it as depending on some peculiar anatomical or physiological condition of the lymphatic system, which renders it specially prone to react under septic agencies. Both the lung and also the omentum of the guinea-pig are largely endowed with lymphatic structures—more so, I think, than are to be found in these organs in man; but positive proof of this kind is very difficult of attainment.

The fact of special aptitude for the action of poisons in particular individuals remains, however, a very positive one. It is shown by the escape of some of the inoculated animals when the same material was used on more than one; and the effect of this predisposition in some cases of other poisons which affect the lymphatic tissues, is well illustrated in the case of typhoid fever—a disease which especially attacks the young, while those past middle life have a greater comparative immunity. The explanation of this has long been taught at University College, by Sir William Jenner, to depend on the progressively diminishing nutritive activity of the follicular apparatus of the intestines and spleen with advancing life, rendering these structures less liable to be influenced by this specific poison.

The relative susceptibility of different genera of animals to this class of agencies has not yet been the subject of any extended experiments. Those hitherto made have been principally conducted with tuberculous matters, but without any very positive results. In one case only is general tuberculosis in a dog recorded by M. Rostan as a result of this proceeding. In other cases, as in a lamb and a ram by M. Colin, tubercles were found only in the lungs; and the same result was twice observed by M. Villemin in cats. M. Lebert also once observed granulations in the lungs of a dog which was the subject of a biliary fistula, and in which he injected phosphoretted oil into the rectum. But local growths in the lungs alone, though possibly tuberculous in their nature, do not, as I have before stated, afford the conclusive proof which I think that we at present require, and which is yielded by the multiplicity of the affection in many organs. In other animals, as in a goat, a cock, and a wood-pigeon (Villemin), or in a crow, cat, and fowl (Vogel), the experiments hitherto conducted have been without results. It would appear desirable that these should be conducted on a yet more extended basis.

I would, however, finally venture to observe that I think we should be cautious in drawing conclusions that the method of the production of tubercle, now partially elucidated, is the sole mode of origin of this disease.

M. Villemin's experiments have, like many others in science, led to a different conclusion from that which the first observer drew. These results appear to me to have a wider scope and to open questions of deeper interest than that of the specificity of tubercle, important as such a demonstration, could it be justified by the facts, would have proved. The subject now rests on a wider physiological and pathological basis respecting its nature and origin. It will doubtless be still a question in some minds whether the results of these experiments are truly tubercle. That question must be settled by the concurrence of other observers, though already there is a wide uniformity of opinion in the affirmative on this point. For my own part, I can only express my conviction, to which I have arrived by a long and careful comparative series of observations, that these growths are identical with the typical forms of tubercle.

I believe, however, that further efforts, which may be directed to the interpretation of these results and to the explanation of the origin of tubercle under these new and hitherto strange circumstances, cannot fail to be productive of new truth. It may yet be long before they receive their full and adequate explanation—an explanation for which further advances in anatomy and in pathological physiology appear to be requisite in order to reveal to us the reason for the peculiar localisation and circumscription of these growths, and to explain the conditions under which they occur. It would appear that but a small tract of a wide field for inquiry has as yet been opened for new researches, though I feel convinced that it is one which will repay the energies of the most careful and attentive observers. The explanations which we may now devise to ourselves are only at present of the nature of more or less tenable hypotheses. I cannot, however, but believe that the new series of facts now disclosed will be proved to be no unimportant portion in the phenomena which must be embraced by these hypotheses, and that in proportion as the nature and mode of origin of tubercle are more fully explored, we may, I think, indulge in the not unphilosophical hope that with the discovery of its causes we may find means at least for its prevention, if not in all cases for its cure.

APPENDIX.

Note to P. 28.

Since the lecture was delivered, I have had proof that even simpler irritants may serve as the starting-point of the tubercular process in the guinea-pig. Of twenty-four of these animals (not included in the preceding numbers), in whom, for purposes of identification in another experiment, I had inserted a small piece of silver suture wire in the nucha, three became tuberculous. The appearances around the seat of injury, together with the implication of the neighbouring lymphatics, left no doubt that the origin of the disease was from this local cause. The escape of all the other animals points to the slighter efficacy of such a simple irritant in producing tubercle, as compared with those previously employed. Access of air to the subcutaneous tissue was of course possible, and thus the question of a "septic" agency must still be entertained; but the possibility nevertheless remains, whether this effect may not be produced by simpler causes of irritation.

R. CLAY, SONS, AND TAYLOR, PRINTERS, BREAD STREET HILL.